Friedel-Crafts Alkylation of Aromatics with Alcohols and Transalkylation Reactions: From Solid Acid Catalysts to Green Ionic Liquids

Sreedevi Upadhyayula
Gul Afreen

ELIVA PRESS

ELIVA PRESS

Sreedevi Upadhyayula

Gul Afreen

The Friedel–Crafts alkylation is a powerful substitution reaction to synthesize alkylphenols. Substitution by an alkyl group results from attack on the aromatic hydrocarbon by a cation (carbonium ion), a neutral fragment (free radical) or an anion (carbanion). The use of alcohol as alkylating agent for Friedel-Crafts reaction gives the benefit of a long stable life of the catalyst and ease of storage and transportation. The alkylphenols prepared from the alkylation of aromatics with alcohol has industrial applications in the production of high-octane gasoline and synthetic rubbers, fibres, detergents, and plastics. The yield of desired alkylphenols is further enhanced by transalkylation of the side products, di-alkylphenols, with similar or dissimilar molecules. This review details the progress of alkylation and transalkylation of aromatics with alcohols over tailor-make zeolite molecular sieves and room temperature ionic liquids to get a high yield and selectivity for a particular reaction of interest.

Published: Eliva Press SRL
Address: MD-2060, bd.Cuza-Voda, 1/4, of. 21 Chişinău, Republica
Moldova
Email: info@elivapress.com
Website: www.elivapress.com

ISBN: 978-1-63648-026-8

A review on Friedel-Crafts alkylation of aromatics with alcohols and transalkylation reactions: From solid acid catalysts to green ionic liquids

Gul Afreen[1], Sreedevi Upadhyayula[1,*]

[1]*Department of Chemical Engineering, Indian Institute of Technology Delhi, Hauz Khas, New Delhi-110016, India*

Corresponding author Email ID: sreedevi@chemical.iitd.ac.in

ABSTRACT

The Friedel–Crafts alkylation is a powerful substitution reaction to synthesize alkylphenols. Substitution by an alkyl group results from attack on the aromatic hydrocarbon by a cation (carbonium ion), a neutral fragment (free radical) or an anion (carbanion). The use of alcohol as alkylating agent for Friedel-Crafts reaction gives the benefit of a long stable life of the catalyst and ease of storage and transportation. The alkylphenols prepared from the alkylation of aromatics with alcohol has industrial applications in the production of high-octane gasoline and synthetic rubbers, fibres, detergents, and plastics. The yield of desired alkylphenols is further enhanced by transalkylation of the side products, di-alkylphenols, with similar or dissimilar molecules. This review details the progress of alkylation and transalkylation of aromatics with alcohols over tailor-make zeolite molecular sieves and room temperature ionic liquids to get a high yield and selectivity for a particular reaction of interest.

Keywords: Alkylation, Transalkylation, Aromatic compounds, Solid acid catalysts, Ionic liquids

Table of Contents

1. INTRODUCTION

Alkylation is the substitution or addition of any alkyl (or aralkyl) group into an organic compound. While the number of possible different alkyl groups is very large, the following are the principal ones of technical importance: methyl, ethyl, propyl, butyl, amyl and hexyl. Substitution by an alkyl group can result from attack on the aromatic hydrocarbon by a cation (carbonium ion), a neutral fragment (free radical) or an anion (carbanion). The importance of alkylation in organic preparations was realized as early as 1877 by Charles Friedel and James Mason Crafts to produce amyl benzene, with the reaction of benzene and amyl chloride [1]. This was apparently the first typical alkylation reaction and came to be known as the Friedel-Crafts alkylation reaction. Friedel-Crafts reactions now find a number of industrial applications in the production of high-octane gasoline, perfumery, synthetic rubber, lubricants, plastics, fibers, and detergents, among others. Various catalysts have been and are being used for alkylation of aromatics starting from original Friedel-Crafts catalyst, AlCl$_3$ to zeolites of the present [2-9]. More recent trend is to tailor-make the zeolite molecular sieves or room temperature ionic liquids to get a high yield and selectivity for a particular reaction of interest.

The discovery of Friedel-Crafts reaction was made with alkyl halide as the alkylating species. However, alkenes came into prominence as alkylating agents later, followed by alcohols in recent years. The use of alcohol as an alkylating agent has some advantages. For example, the deactivation rate is observed to be lower in case of alcohol as alkylating agent. Moreover, choice of alkenes (which are in gas phase) as alkylating agents is usually associated with high storage and transportation costs. The direct use of ethyl alcohol for the synthesis of ethylbenzene is again economically significant in countries like India where the alcohols are obtained as efficient raw materials from the biomass and contribute in the formation of chemicals. In cumene manufacture, often the propylene feed contains trace levels of cyclopropane, which increase the yield of the highly undesired byproduct, normal propyl benzene (n-PB).

During the aromatics alkylation, dialkylated side products always tag along with the monoalkylated products. The yield of monoalkylated products can be increased by transfer of an alkyl group from the dialkylated product to nearby similar/ dissimilar compound by a process called transalkylation. The transalkylation is an industrially significant and performed generally at high temperatures and pressures over solid acid catalysts. In this review, a thorough study on the alkylation and transalkylation of aromatics with alcohol over solid acid catalysts is reported. The review further mentions the drawbacks of solid acid catalysts and extends the usage of green ionic liquids as solvent and catalyst in this alkylation reaction. An elaborated description of the reaction mechanism and the role of catalysts in the reaction is explained. The industries involved in such alkylation reactions is also summarized here to understand the importance of this reaction in commercial applications.

3

2. ALKYLATION OF AROMATICS WITH ALCOHOLS OVER SOLID ACID CATALYSTS

2.1. Alkylation reactions of aromatic hydrocarbons

The specific attachment of an alkyl group to an aromatic hydrocarbon is of importance because of the varied applications of the products and their consequent large volume of manufacture. Substitution by an alkyl group can result from attack on the aromatic hydrocarbon by a cation (carbonium ion), a neutral fragment (free radical), or an anion (carbanion). Friedel-Crafts alkylations fall under the category of nuclear alkylation involving the substitution of an aromatic hydrogen. This is carbon-to-carbon alkylation (also known as C-alkylation) wherein the carbon of the alkyl is bound to carbon of the aromatic compound. The Friedel-Crafts alkylation reaction is also often accompanied by polyalkylation. Since alkyl groups activate the aromatic ring towards further attack, there is a marked tendency for polysubstitution during the alkylation. This affects the yield of the monoalkyl product. However, the alkyl groups can be transferred from one aromatic ring to the other using suitable catalysts. This is known as transalkylation (or disproportionation). This reaction can result in the formation of high value monoalkylated products with high demand from the lesser important polyalkylated compounds. For example, transalkylation of diethylbenzenes to ethylbenzene is an industrially viable transalkylation reaction. In Friedel-Crafts alkylation, the alkylating agent and the catalyst such as aluminium chloride and hydrogen chloride, react to form either carbonium ion or complex, which further reacts with the aromatic ring [2]. Thus, the reactions occuring in the alkylation of olefins is shown in **Figure 1**.

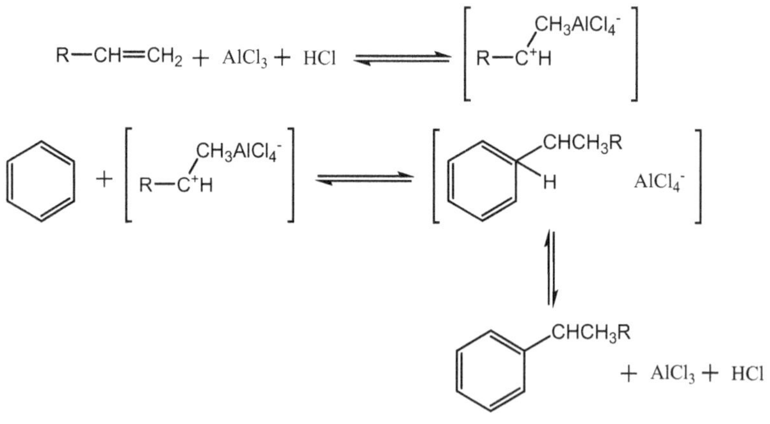

Figure 1. Alkylation reaction of olefins.

4

The aromatic ring to which the olefin gets attached may be that of benzene, substituted benzene or more complicated ring systems like naphthalene or anthracene. Friedel-Crafts reactions are complicated by the rearrangement of the attacking agent and in some cases, the aromatic starting materials [2]. There is a tendency for a carbonium ion formed during the reaction, to rearrange to a stable secondary or tertiary carbonium ion. Kaeding et al. [10] suggested a mechanism for aromatic alkylation with alkylating agents other than olefins over zeolite molecular sieve catalysts (solid acid catalysts) in agreement with the report by Anderson et al. [11]. They showed that the methyl groups remain intact while hydrogens on the aromatic ring are easily exchanged under alkylation conditions as shown in **Figure 2**.

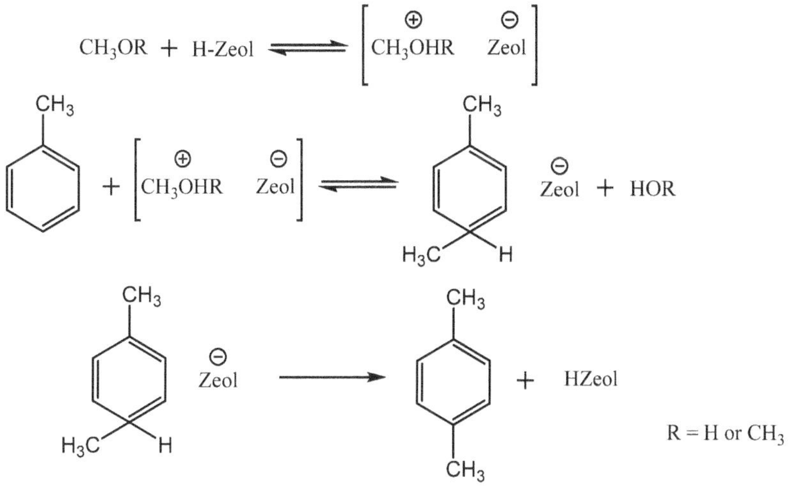

Figure 2. Alkylation of aromatic with alkylating agents other than olefins over zeolite molecular sieve catalysts.

Butylated phenols are produced by alkylating phenol with various phenol compounds such as, phenol, cresol, xylenol and cenitol etc. based on industrial application. Tertiary butylation of phenol is of industrial interest due to the products formed and is therefore widely studied. 2-*tert*-butyl phenol is a precursor in the industrial production of agrochemicals and antioxidants. Triphosphate and benzotriazole derived from 2,4-di-*tert*-butylphenol are used as co-stabilizers for polyvinyl chloride and UV absorbers in polyolefins, respectively. The selectivity to products mainly depends on the catalyst acidity and reaction parameters. The selectivity to various products based on the acidity of the catalyst is shown in **Figure 3** [12]. Weakly acidic catalysts and hydroxyl group (-OH) present on the aromatic ring kinetically favors O-alkylated product (*tert*-butylphenyl ether). Moderately acidic catalysts favor both O- and C-alkylation. Even the thermodynamically unfavored

2-isomer promptly isomerizes to 4-isomer due to steric hindrance. Strong acid catalysts produce more di-alkylated products. At higher acidity and temperature, the reaction produces 3-*tert*-butyl phenol. To reduce oligomers and O-alkylated products, the catalyst and enantioselectivity to C-alkylated products needs to be improved.

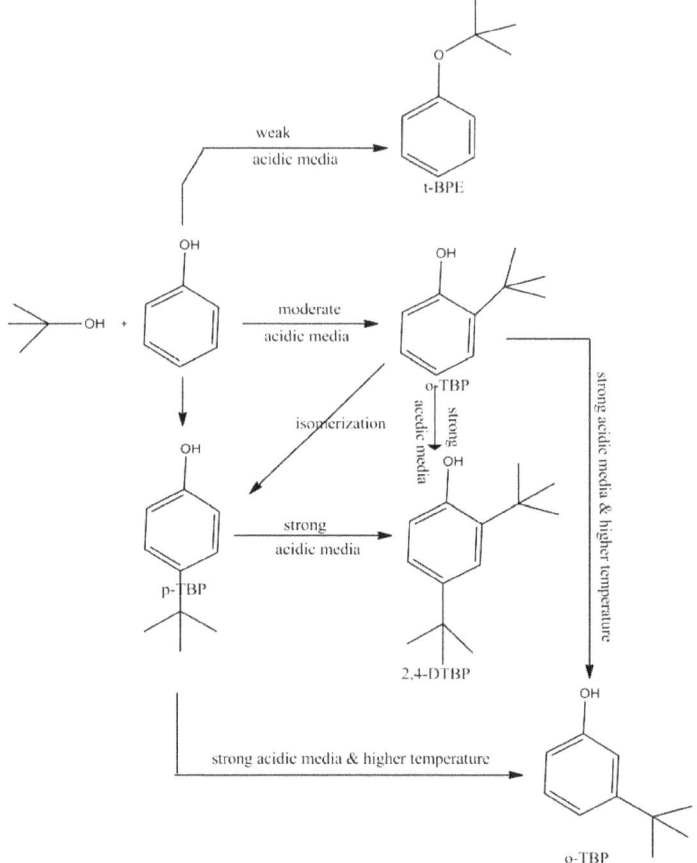

Figure 3. Alkylation of phenol with *tert*-butanol in different strength of acidic media.

Tert-butylation of *p*-cresol produces monoalkylated product, 2-*tert*-butyl-*p*-cresol and di-alkylated product, 2,6-di-*tert*-butyl-*p*-cresol which are combinedly known as Butylated hydroxytoluene (BHT) [13,14]. BHT is used as an antioxidant. It is also used as a raw material in the synthesis of oil-soluble phenol-formaldehyde resins [13]. It is industrially manufactured by alkylation of *p*-cresol with isobutylene over acid catalysts as shown in **Figure 4**.

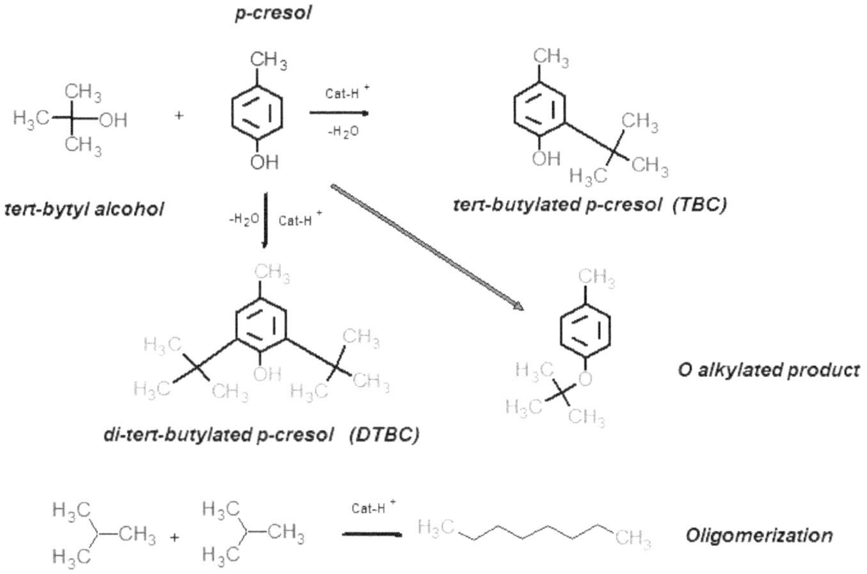

p-cresol

tert-bytyl alcohol

tert-butylated p-cresol (TBC)

di-tert-butylated p-cresol (DTBC)

O alkylated product

Oligomerization

Cat-H⁺
-H₂O

-H₂O   Cat-H⁺

Cat-H⁺

Figure 4. Reaction mechanism of *p*-cresol alkylation with *tert*-butanol over acidic catalyst.

The process route for production of butylated phenols using isobutylene as the alkylating chemical has some inherent problems like difficulty in availability, handling, and transportation of *iso*butylene, especially during the production of fine and specialty chemicals in smaller amount. Hence, an attractive alternate route is to replace *iso*butylene with *tert*-butanol or methyl tertiary butyl ether (MTBE) to overcome these problems. Pure *iso*butylene is generated *in-situ* from *tert*-butanol or MTBE, and the by-product methanol generated in case of MTBE is a widely used raw material in chemical plants [14-16]. Unfortunately, the methanol further participates in the reaction and adversely affects the product yield and selectivity. The propene oxide synthesis from ARCO process releases *tert*-butanol as byproduct. Dehydration of *tert*-butanol to *iso*butylene yields only water as a side-product, which does not affect the product yields and selectivity adversely.

2.2. Transalkylation reactions of aromatic hydrocarbons

As mentioned earlier, the Friedel-Crafts alkylation is also often accompanied by polyalkylation affecting the yield of the monoalkylated product. It was shown that alkyl groups could be transferred from one aromatic ring to another by the catalytic effect of aluminium chloride and hydrogen chloride [2]. For example, when ethylbenzene was heated with the catalyst, a mixture of benzene, ethylbenzene, diethylbenzene (mainly *m*- and *p*- isomers), and higher boiling materials were produced (**Figure 5**).

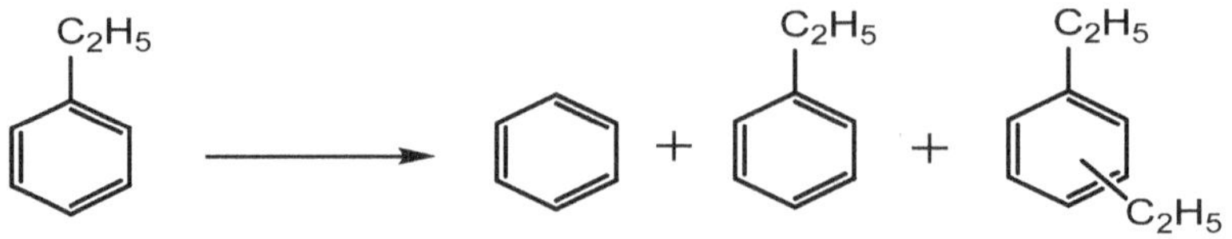

Figure 5. Transalkylation reaction of ethylbenzene.

This reaction is of industrial importance as high value monosubstituted chemicals are formed from the low value polysubstituted products. Earlier, these disproportionation or transalkylation reactions were carried out using Friedel-Crafts catalysts, however, the reactions like disproportionation of toluene and formation of xylenes are recently performed over ZSM5 zeolites developed by Mobil Company [17]. Information available about possible reaction mechanism is limited. Three different reaction mechanisms are reported so far. The first step of all these mechanisms is the reactant adsorption on the active acidic site of the catalyst, followed by generation of carbonium ion formation. According to the first type of mechanism, transalkylation takes place with a methyl transfer from the carbonium ion, to a nearby benzene ring as shown in **Figure 6(a)** [18]. In the second type of mechanism, dealkylation of carbonium ion takes place followed by alkylation as shown in **Figure 6(b)** [18]. The secondary or tertiary carbonium ions are stable and increases the probability of this mechanism. In third mechanism it is suggested that transalkylation reaction involves a 1,1-diphenyl alkane type of intermediate as shown in **Figure 6(c)** [19]. In recent years, transalkylation reactions of diethylbenzene (DEB) with benzene to form ethylbenzene and di*iso*propylbenzene (DIPB) with benzene to form cumene has been reported [20-27].

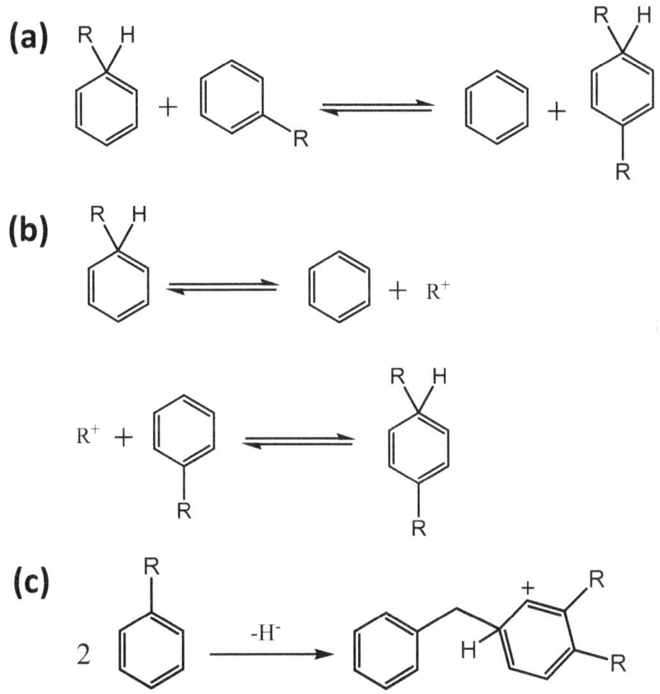

Figure 6. (a) Transalkylation reaction mechanism with a methyl transfer from the carbonium ion to a benzene ring, **(b)** Dealkylation of carbonium ion followed by alkylation, **(c)** Transalkylation reaction involving a 1,1-diphenyl alkane intermediate.

2.3. Catalysts for alkylation and transalkylation reactions

Friedel-Crafts alkylation or transalkylation is both Brönsted and Lewis acid catalyzed reactions [2,3]. For the liquid phase alkylation reaction, anhydrous aluminium chloride is preferred as a catalyst, although a co-catalyst or promoter like hydrogen chloride is usually needed to obtain efficient alkylation. In the vapour phase alkylation, the most commonly used catalyst is phosphoric acid supported on keiselgurh (also known as solid phosphoric acid (SPA) catalyst). Phosphoric acid effectively alkylate benzene to ethylbenzene, *iso*propylbenzene, etc in the presence of by-product water. Several other catalysts have also been reported for the aromatics alkylation such as Lewis acidic metal halides (AlCl$_3$.CH$_3$NO$_2$, BF$_3$, TiCl$_4$, FeCl$_3$, SnCl$_4$, ZnCl$_2$), Bronsted acidic protonic acids (HF, H$_2$SO$_4$, Nafion-H, polyphosphoric acid), and inorganic oxides (phosphorous pentoxide on alumina) [2-4]. Most of these catalysts are in liquid phase and their use entails problems of handling, safety, equipment corrosion, toxicity, and waste disposal. Hence, heterogeneous catalysts can suitably replace these homogeneous catalysts where gas phase is advantageous in the process technology [5]. Solid acid catalysts like molecular sieves are eco-friendly and corrosion-free. Also, the combination

9

of acidity and shape selectivity of molecular sieves makes them potential catalysts for various alkylation and transalkylation reactions. Some of the important and interesting organic reactions, which were not feasible because of low activity, selectivity and/or service life of the catalyst, can now be commercially exploited using molecular sieve catalysts. The zeolite molecular sieves is extensively used to partially replace sodium thiophosphate builders and water softeners from low-phosphate detergents [28]. Besides, they are also utilized widely in treatment plants of waste water and agricultural waste [29].

The catalytic application of molecular sieves in the Petroleum and Petrochemical Industry is their most vital application. Catalytic cracking processes use 10-40% Faujasite Y zeolite dispersed amorphous support like clay or SiO_2-Al_2O_3. The "Parex Process" of M/s UOP also uses Faujasite zeolites in the isolation of p-xylene from its isomers and ethyl benzene. Also, Lindox and Unox processes use low silica zeolites in their pressure swing adsorption. Other petrochemical processes using these catalysts are Mobil-Badger ethylbenzene process (from benzene and ethyl alcohol), ALBENE one step ethylbenzene process also from benzene ethyl alcohol, Mobil's methyl alcohol to gasoline (MTG) process, alkylation between toluene and methyl alcohol to form p-xylene, etc.

2.3.1. Zeolite molecular sieves as alkylation and transalkylation catalysts

Zeolite molecular sieves are hydrated, crystalline, microporous, three dimensional aluminosilicates framework consisting of SiO_4 and AlO_4 building blocks [30] wherein Si and Al are the tetrahedral atoms (T atoms) sharing the adjacent oxygen. The empirical formula to represent a unit cell of such zeolites can be written as follows [31]:

$$M_{x/n}[(AlO_2)_x(SiO_2)_y]zH_2O$$

The extra negative charge due to aluminium is compensated by cations (say M) of n valency such as Na^+, NH_4^+ or H^+. The total T sites is calculated as x+y and the ratio of y:x (in general >1) is a controlling factor of acidity and structure of the zeolite.

The significant applications of zeolites as catalysts and adsorbents are based on the porosity defined by the T atoms [30,31]. The pores in the zeolitic framework are classified in terms of rings as: 8 membered (small pore), 10 membered (medium pore), and 12 membered (large pore) ring. The dimensions and contours of the pores are dependent on the factors such as conformation based on relative T and O atoms, Si/Al ratio, cationic size and position, temperature, and framework structure of the zeolites [31].

The reactivity of zeolites is based on the number of active sites created by the charge imbalance due to the aluminium atoms [32]. Classical Brönsted and Lewis acid models have been used to classify active sites in zeolites. Brönsted acidity arises in the zeolite when the cation balancing the anionic framework charge is a proton (H^+). Differences in the acidic strength between different zeolites is often related to T-O-T bond lengths and bond angles and crystal resonance energy [31].

A trigonally co-ordinated aluminium atom, which acts as an electron acceptor, behaves as a Lewis acid. Higher temperature (>500°C) can result in the protonic Brönsted acid sites turning to Lewis acid sites by dehydroxylation [11]. Dealumination by hydrothermal treatment of zeolites has also been found to produce a variety of cations and neutral species which function as Lewis acids [33]. These cations also induce activity on nearby Brönsted acid sites. Guo et al. reported the benzene alkylation with a mixture of ethane and propene over H-ZSM-5 zeolite [34]. They synthesized the aromatic hydrocarbons in C_8-C_{15} range with 55% yield. They described the advantage of reducing the catalyst particle size in the nanosize range by rapid mass diffusion as well as increasing the Al amount to increase the acidity on the enhanced yield of the desired range of products.

2.3.2. Aluminophosphate molecular sieves (AlPO₄) as alkylation and transalkylation catalysts

Aluminophosphates ($AlPO_4$s) are recognized as microporous crystalline oxides structurally analogous to zeolites [35-38]. Unlike zeolites, these materials have no ion-exchange capacity as they possess a balanced framework of charges. In $AlPO_4$s, the framework sites are occupied by Al^{3+} or P^{5+}. The average of the ionic radii of Al^{3+} and P^{5+} is 0.28 Å which is similar to the ionic radii of Si^{4+}(0.26 Å). The Al atoms are present in co-ordination in a neutral framework. Structural diversity is found in $AlPO_4$ materials even with such limited variation in chemical composition. The overall composition is written as:

$$x R.Al_2O_3.1.0(+0.2)P_2O_5.y H_2O$$

where, R is either $(NH_4)^+$ or an organic amine. The x and y are moles of R and water present in the pores of the framework. $AlPO_4$s also obey the Lowenstein's rule, according to which (a) no two central atoms of the two adjacent tetrahedra joined by oxygen bridge can be aluminium, the other atom must be other than aluminium with an ionic electrovalency of ≥4, (b) an oxygen ion harbors two Al ions only when atleast one of the Al ions must have co-ordination number >4. $AlPO_4$s contain alternating tetrahedra of AlO_2^- and PO_2^+. These structures exhibit only 4-, 6-, 8- and 12-membered rings. Other ring sizes are observed only when structural alumina has a co-ordination number greater than 4.

Structure of $AlPO_4$-5 contains alternating Al and P atoms forming 4- and 6- membered rings. The pore system is a channel created by 12-membered rings. It has hexagonal symmetry with unit cell parameters a = 13.73 Å, c = 8.48 Å and γ = 120°. $AlPO_4$-11 contains 1-D 10-membered ring channels consisting of 6.7 x 4.4 Å pore size openings. It has orthorhombic symmetry made with lattice parameters of a = 13.5 Å, b = 18.5 Å, and c = 8.4 Å [36]. Due to the balanced framework of charges, the $AlPO_4$ molecular sieves are neutral in charge and catalytically inactive. Hence, to make these molecular sieves active for reactions requiring acidic catalysts, silicon is substituted into their framework forming Silicoaluminophosphate, i.e., SAPO-n ('n' representing the structure type) molecular sieves. Theoretically, silicon can substitute for both aluminium and phosphorous. The substitution of silicon can be in three ways as follows:

(a) Silicon replacing aluminium to give positive framework and anion exchange properties.

(b) Silicon replacing phosphorous to give an anionic framework like that of zeolites.

(c) Both aluminium and phosphorous are simultaneously substituted by a pair of silicon atoms with no net framework charge.

So far, no observations have been made on substitution of silicon for aluminium, while the other two mechanisms (a&b) are found to have occurred in case of SAPO-n [31]. Although the SAPO-n molecular sieves have similar pore structures as those of zeolites, they are comparatively weaker in acidity owing to the presence of phosphorous [39-42]. The combined opportunity of tuning of porosity and somewhat acidity recommends the use of SAPO-n ('n' represents the structure type) sieves in shape selective catalytic applications including alkylation and transalkylations [42-46]. SAPO-5 is made of 12 membered non-intersecting ring with pore diameter of 7.3 Å creating an aluminophosphate-5 topology [31].On the other hand, SAPO-11 is composed of aluminophosphate-11 topology with 10 membered non-intersecting ring having pores in the range of 3.9-6.3 Å. Like zeolites, the Brönsted acidity of the AlPO$_4$ based molecular sieves depends on both framework composition and structure type. Unlike the zeolites, the latter exhibit a high level of compositional as well as structural diversity. Among the large pore structures, SAPO-5 exhibits strong acidity [46]. In SAPO-5, the relationship between silicon substitution and acidic site generation is about 1:1 up to 1% substitution. Above this, the acid site production comes down from the 1:1 ratio. At 6% silicon substitution, only around 3% of silicon forms Brönsted acid sites [47].

2.3.3. Catalysts for tert-butylated phenols

Several homogeneous and heterogeneous phase catalysts are reported for *tert*-butylation of phenols. Some of the examples are Lewis acidic AlCl$_3$ and BF$_3$ [48], Bronsted acidic HF, H$_3$PO$_3$, H$_2$SO$_4$, etc [49], cation-exchanged resins [50-52], zeolites [53,54], unsupported and supported mesoporous materials [55,56], heteropoly acids, water in the near critical and super critical temperature [57]. Though cation-exchanged resin showed satisfactory performance, yet they have issues like resin fouling and thermal instability. Alkylation of phenols using sulfuric acid catalyst was studied in batch and bubble column by Sharma et al. [58]. Solubility of isobutene in sulfuric acid and alkylation reaction kinetics were studied. Sharma and researchers also reported the alkylation over cation-exchanged resins [59]. Calcined magnesium-aluminium hydrotalcites was used for the alkylation of phenol with *iso*-butanol at temperatures between 350 and 500°C by Padmasri et al. [60]. Mainly, the products o-*tert*-butyl phenol and 2-*tert*-butyl phenol were formed along with some o-butenyl phenol and 2-butenyl phenols. Zhang et al. checked the catalytic performance of Hβ zeolite towards this reaction [61]. They observed that weak, medium and strong active acidic sites on the zeolite Hβ favors the formation of *o-tert*butyl phenol, *p-tert*butyl phenol, and 2,4-di-*tert*butyl phenol, respectively. Optimization of reaction temperature to ~145 °C, reactant molar ratio, and acidity to moderate was found to further enhance the *p-tert*butyl phenol selectivity. Satyanarayana et al. showed the role of silicon incorporation on AlPO$_4$-t-11, 31, and 41 structures of medium pore size on phenol butylation [62]. Zhang et al. screened several zeolites for this reaction and observed zeolite HY to be the best in temperature range of 125-165 °C [63].

Tang et al. investigated the temperature dependence on the selectivity of products in phenol alkylation with methyl-*tert*-butyl ether (MTBE) over mesoporous MCM-41 and observed that *p-tert*-butylphenol (*p*-TBP) and 2,4-di-*tert*-butylphenol (2,4-DTBP) is formed preferably at lower temperatures while 2,4-DTBP dealkylated at higher temperature [64]. Ganapati et al. screened various montmorillonite K-10 clays derivatives in a batch reactor using both methyl-*tert*-butyl ether (MTBE) and *tert*-butanol alkylating agents for phenol alkylation [65]. They found the order of catalytic performance as: 20% (w/w) dodecatungstophosphoric acid (DTP)/K-10 > K- 10 > 20% ZnCl$_2$/K-10 > Al-exchanged K-10 > Zr-exchanged K-10 > Cr-exchanged K-10. Padmasri et al. compared the catalytic performance of oxides of calcined hydrotalcites in the order Mg-Al > Zn-Al > Mg-Cr in phenol alkylation with *iso*-butanol resulting in the formation of butenyl phenols and butyl phenols in the temperature range of 350-500°C [66]. Dumitriu et al. compared the phenol alkylation with *tert*-butanol in liquid phase over zeolites H-FAU, H-BEA, and H-MOR [53]. The reaction over these catalysts was observed to produce *tert*-butylphenyl ether as the main product. The three dimensional porous framework of H-FAU and H-BEA was found to be more active in the reaction than the single dimensional H-MOR.

Dapurkar et al. reported that H-AlMCM-48 having 3-D pore system showed higher catalytic activity time on stream than hexagonal H-AlMCM-41 in the phenol/*tert*-butanol alkylation [67]. Tang et al. observed weak and medium acidic sites to be active centers for the formation of *p-tert*-butyl phenol and 2,4-di-*tert*-butyl phenol in phenol/ *iso*butylene alkylation over zeolite BEA [68]. Mathew et al. reported phenol/*iso*butene alkylation dependency on reaction parameters like temperature, time, reactant composition, and WHSV [69]. Devassy et al. reported that ZrO$_2$ supported 12-tungstophosphoric acid (15%) showed significant catalytic activity towards *p*-cresol alkylation by *tert*-butanol with *p*-cresol conversion of 61% and selectivity to butylated product of 99.5% at 130 °C temperature and 4 h^{-1} LHSV [70]. Karthik et al. investigated the reaction of *m*-cresol with *tert*-butyl alcohol over hexagonal mesoporous AlPO, MgAPO, and CoAPO at 175-325 °C [71]. The ortho substituted 2-*tert*-butyl-5-methyl phenol was formed in major amount whereas, di-*tert*-butyl ether was also formed in small amount. Vinu et al. investigated phenol/*tert*-butanol alkylation over Fe-Al-

MCM-41 (Si/Al=40; Si/Fe=40) and observed a phenol conversion of 70% with selectivity for *p-tert*-butylphenol of 75% at 200 °C [72].

Shinde et al. reported alkylation of phenol and cresols with *tert*-butanol over Fe-modified montmorillonite K-10 in liquid phase with about 66% *para*-alkylated product selectivity [73]. Nandini et al. performed vapour phase alkylation between phenol and *tert*-butanol using 10, 20 and 40 wt% phosphotungstic acid supported on aluminophosphate as catalysts getting around 57% aromatic conversion at temperatures 190–250 ˚C [74]. Dumitriu et al. observed no enhancement in yield of *p-tert*-butyl phenol during phenol/*tert*-butanol alkylation over MWW-based MCM-36, MCM-22, and ITQ-2 catalysts [75]. Vinu et al. used ordered mesoporous Al-SBA-15 (Si/Al=45) for this reaction to obtain 86% conversion at 150 °C [76]. Devassy et al. studied this alkylation reaction over 15% silicotungstic acid (STA) modified zirconia [77]. Phenol was converted upto 95% into 4% *o-tert*-butyl phenol, 59% *p-tert*-butyl phenol, and 35% 2,4-di-*tert*-butyl phenol at the optimum of 140 °C, 1:2 phenol/*tert*-butanol mole ratio, and 4 h^{-1} LHSV. Sarish et al. also optimized the reaction parameters to 130 °C temperature, 3:1 mole ratio of *tert*-butanol/*p*-cresol, and 10 ml h^{-1} flow rate to achieve a *p*-cresol conversion of 69.8% and selectivity to products as: 92% *o-tert*-butyl-*p*-cresol, 6% 2,6-di-*tert*-butyl-*p*-cresol, and ~1% *p*-cresol-*tert*-butyl ether using WO$_x$/ZrO$_2$ catalyst in continuous flow reactor [78]. Ojha et al. reported kinetics of this reaction performed in a batch reactor in the presence of a zeolite prepared from fly ash. However, they achieved the phenol conversion of only 28% at 70 ˚C [54]. Wu et al. compared the catalytic activity of Al-SBA-15 with Al-MCM-41 and observed that phenol was converted to 75% with 2,4-DTBP selectivity to 31% in case of Al-SBA-15 as opposed to 61% conversion and 13% selectivity in case of Al-MCM-41 at 145°C [79]. Yadav et al. prepared sulphated ZrO$_2$ and modified it to superacidic catalysts namely, UDCaT-4, UDCaT-5, and UDCaT-6 to use them in the *m*-cresol alkylation with *tert*-butanol [80]. Among all the catalysts, they reported the maximum *m*-cresol conversion of 89% and overall selectivity to mono-C-alkylated products of 90% over 0.03 g/cm^3 UDCaT-5 at 120 °C and *m*-cresol/*tert*-butano mole ratio of 3:1. Devassy et al. investigated phenol/*tert*-butanol alkylation over 12-molybdophosphoric acid supported on ZrO$_2$ and reported 80.6% phenol conversion with selectivity for 2-*tert*-butylphenol, 4-*tert*-butyl phenol, and 2,4-di-*tert*-butyl phenol to be 11.5%, 25.7%, and 55.2%, respectively [81].

Duan et al. prepared and characterized 2-methylpyridinium based protic ionic liquids to test their catalytic properties in phenol *tert*-butylation [82]. A 95% phenol conversion with 1%, 10%, 82% and 7% selectivities to 2-*tert*-butylphenol, 4-*tert*-butyl phenol, 2,4-di-*tert*-butyl phenol, and 2,6-di-*tert*-butyl phenol, respectively, was obtained with 2-methylpyridinium trifluoromethanesulfonate ([2-MPyH]OTf) at 120 ˚C. Kurian et al. exchanged transition metal on pillared montmorillonites and obtained 54% phenol conversion and 96% 4-*tert*-butyl phenol selectivity during phenol/*tert*-butanol alkylation [83]. Kamalakar et al. used supercritical CO$_2$ medium for cresol/*tert*-butanol alkylation over MCM-41 supported tungstophosphoric acid which yielded 58% 2,6-di-*tert*-butyl-4-methylphenol (2,6-DTBPC) at 110 °C [84]. On the other hand, titania supported 12-tungstophosphoric acid yielded 73% 2-*tert*-butyl cresol during *p*-cresol/*tert*-butanol alkylation at 130 °C as reported by Kumbar et al. [85]. In another report, Nandini et al. prepared Al-MCM-41 (Si/Al=20) supported phosphotungstic acid (20%) for phenol/*tert*-butanol alkylation in vapor phase at 190–275 °C [86]. They obtained 89% phenol conversion at 190 °C with 84% 4-*tert*-butyl phenol selectivity, 6% 2-*tert*-butyl phenol selectivity, and 10% 2,4-di-*tert*-butyl phenol selectivity.

Kumar et al. reported that SBA-15 supported by 30% phosphotungstic acid achieved 70% phenol conversion at 190 °C and 72% 4-*tert*-butyl phenol selectivity [87]. Tang et al. compared the alkylating agents *iso*butylene, MTBE and *tert*-butanol as well as different zeolite catalysts in phenol alkylation and identified the reactivity in order of: *iso*butylene > MTBE > TBA for alkylating agents and H-β > H-Y > H-ZSM-5 ≈ H-Al-MCM-41 for catalysts [88]. Selvaraj et al. reported the effect of different Si/Al ratio of Al-MCM-41 on the reactivity of *p*-cresol/*tert*-butanol alkylation reaction [89]. A Si/Al ratio of 21 was found to be highly reactive giving 88% *p*-cresol conversion and 90% 2-*tert*-butyl-*p*-cresol selectivity at alcohol:cresol mole ratio of 2:1 and 90 °C temperature [90]. Ng et al. used the sulphation method to generate Bronsted acid sites in Al-MCM-41 and observed that this modified catalyst was active towards *tert*-butanol/phenol alkylation [91]. Kumar et al. examined MCM-22 (Si/Al~24–76) as a catalyst for *tert*-butanol/phenol alkylation and attained 85% phenol conversion and 76% *p*-*tert*-butyl phenol selectivity at 175 °C [92]. Reddy et al. reported that in the presence of promoters MoO_4^{2-} and WO_4^{2-} used with ZrO_2, good catalytic performance for phenol/*tert*-butanol alkylation is attained [93]. Meloni et al. alkylated phenol with *tert*-butanol in liquid phase over MWW based catalysts MCM-22, MCM-36, and ITQ-2 [94]. They concluded that the reaction pathway involved many parallel and/or successive steps, the main reactions being O-alkylation and C-alkylation. Xu et al. investigated the hierarchical ZSM-5 zeolite in *tert*-butylation of phenol with *tert*-butanol and acquired 81% aromatics conversion and 41% 2,4-di-*tert*-butyl phenol selectivity at 149 °C [95]. Bhatt et al. investigated liquid phase phenols and cresols alkylation over 30% 12-tungstophosphoricacid supported onto zirconia where 100% phenol was converted to 97% *p*-isomers selectivity at 80 °C for 6 h and aromatic/alkylating agent mole ratio of 10 [56]. Priya et al. exchanged Zn^{2+}, Ce^{3+}, Fe^{3+}, La^{3+} ions on MAPO-36 to carry out *tert*-butanol/phenol alkylation [96]. The maximum aromatic conversion was 50% and selectivity to 4-tert-butyl phenol was 69% at 300 °C, 1:1 feed ratio, and 2.80 h^{-1} WHSV. Bachari et al. reported 75% phenol conversion at 150 °C over gallium loaded hexagonal mesoporous silica that yielded 42% 4-*tert*-butyl phenol and 30% 2,4-di*tert*-butyl phenol [97]. Modrogan et al. investigated which type of catalyst acidity showed better activity in alkylation between phenol and *iso*butene [98]. The used Lewis acidic catalyst (phosphonium ionic liquid immobilized on silica), Bronsted acidic catalyst (Amberlyst 15), and Bronsted/Lewis dual acidic catalyst (WO_3/ZrO_2). The results suggested that Bronsted/Lewis dualilty present in WO_3/ZrO_2 was a suitable active catalyst under moderate reaction conditions.

Liao et al. suggested that protonic COK-5 zeolites (Si/Al=20-30) gives good activity results in phenol *tert*-butylation [99]. Zhou et al. performed quantum chemical simulations in reconciliation with experimental results to determine the reaction mechanism of *m*-cresol alkylation with *tert*-butanol using SO_3H-functionalized ionic liquid [100]. Ghiaci et al. improved Al-MCM-41 with 5-35 wt% H_3PO_4 to use them in the phenol/*tert*-butanol alkylation in vapour phase at 110-220 °C [101].

3. AROMATICS ALKYLATION WITH ALCOHOLS USING IONIC LIQUID (IL) CATALYSTS

Both sulfuric acid and hydrofluoric acid catalysts have disadvantages [102,103]. Anhydrous hydrofluoric acid is highly toxic, corrosive and volatile. Therefore, HF alkylation plants are equipped with expensive safety systems to prevent leakage of the acid into the atmosphere. Sulfuric acid is less toxic compared to HF, also corrosive but much less volatile than HF. The acid consumption is due to formation of heavy hydrocarbons (acid soluble oil), which dissolve in the acid and dilute it. The cost for regenerating the acid (to remove water and heavy hydrocarbons) is about twice or thrice the price of sulfuric acid. Furthermore, the acid consumption is almost $1/3^{rd}$ of the total operation costs. It is very important to mention that large quantities of spent sulfuric acid should be regenerated [104]. Various catalysts have been and are being used for alkylation of aromatics starting from these acids, $AlCl_3$ to zeolites. In both academic and industrial research centers solid acid catalyst, such as Y zeolites [105], sulfated zirconia [106], mesoporous materials [107] and supported poly acid [108] have been studied as possible alternatives to HF and H_2SO_4. Some of these catalysts have replaced the conventional Friedel-Crafts catalysts. However, these catalysts pose problems of deactivation by coking and spent catalyst disposal. Thus, a continuous necessity to develop new economical and environment friendly catalyst systems and alkylation technologies is required.

3.1. Ionic liquids

Ionic liquids (ILs) were introduced as a new class of solvents and reaction media in the last decade [109-139]. ILs are ill coordinated ionic salts which are, unlike regular salts, present in liquid state at temperatures below 100 °C including room temperature ILs. The ill coordination results from the presence of delocalized charges and organic moiety (atleast one in number) that restricts the creation of a stable crystal. The advantages of ILs lie in their low volatility, dipolar property, flexibility towards varying ions, and ability of hetereogenization. The non-volatile nature offers lower toxicity and microwave-friendly synthesis in contrast with volatile solvents. ILs accelerate the reactions due to rapid excitation by microwaves or heating owing to their dipole properties. The important properties like acidity can easily be tuned by varying anions, cations or functional groups present on the cations. The further advantage of such ILs can be implied by introducing a biphasic system rendering the ease of catalyst separation. Moreover, immobilizing ILs on a suitable solid support is beneficial in catalysis aspects.

3.1.1. Cations

Cations for a given anion present in salts are mainly influenced by melting point. A low symmetry of the cations results in low melting points. **Figure 7** summarizes the common cations used in the preparation of ionic liquids. The well characterized salts based on N,N-dialkylimidazolium cations gained special interests as they have broad range of physico-chemical properties. Besides this group, pyrrolidinium, ammonium, pyridinium or lithium actions are of increasing interest for many different applications [140]. Phosphate ionic liquids based on organic polycations (**Figure 8**) are in focus for organic electrochemical processes, and media for "green" chemical reactions. Furthermore, potential applications for electrochemical storage cells are being made [141]. Organic polycations combined with bromide form salts melting at about 70°C [142].

3.1.2. Anions

The anions are usually classified in two groups: The first class contains polynuclear anions, for instance $[Al_2Cl_7]$, $[Al_3Cl_{11}0]$, $[Fe_2Cl_7]$, $[Sb2Cl_{11}]$ or $[Au_2Cl_7]$. These are designed by reacting mononuclear anions like $[AlCl_4]$ with their analogous Lewis acids, e.g. $AlCl_3$. Typical mononuclear anions belonging to the second class are X^-, $[BF_4^-]$, $[CuCl_2^-]$, $[ZnCl_3^-]$, $[SnCl_3^-]$, $[N(SO_2CF_3)_2^-]$, $[PF_6^-]$, and $[CH_3SO_3^-]$ [143].

Figure 9 shows the most commonly used cation and anion based on their miscibility with water. The cationic and the anionic components of ILs can be controlled to alter the polarity of the ILs to obtain hydrophilic and hydrophobic ILs. Several studies show that the anions have substantial effect on the polarity of the ILs [144]. Anions like NTf_2^-, BF_4^-, PF_6^-, OTf^-, etc. leads to form hydrophobic ILs [145]. Furthermore, the viscosity of the ILs can be controlled by altering the cationic chain length and the introduction of suitable anionic counterpart. Certain anionic components like $(NCN)_2^-$, NTf_2^-, OTf^-, etc. are believed to decrease the viscosity of the ILs by decreasing the anion to cation interaction as well as intermolecular interactions [146]. These unique properties of the ILs make them useful compounds in processes suffering from separation issues [147].

3.2. Application of ILs

The distinctive and unique features of the ionic liquids make them capable of widely applicable in various industrial applications [144]. In catalysis, ILs are used widely as catalysts and solvents for various industrially useful reactions [145,148]. Furthermore, ILs have been used rigorously in extraction processes as a substitute to various organic solvents [149].

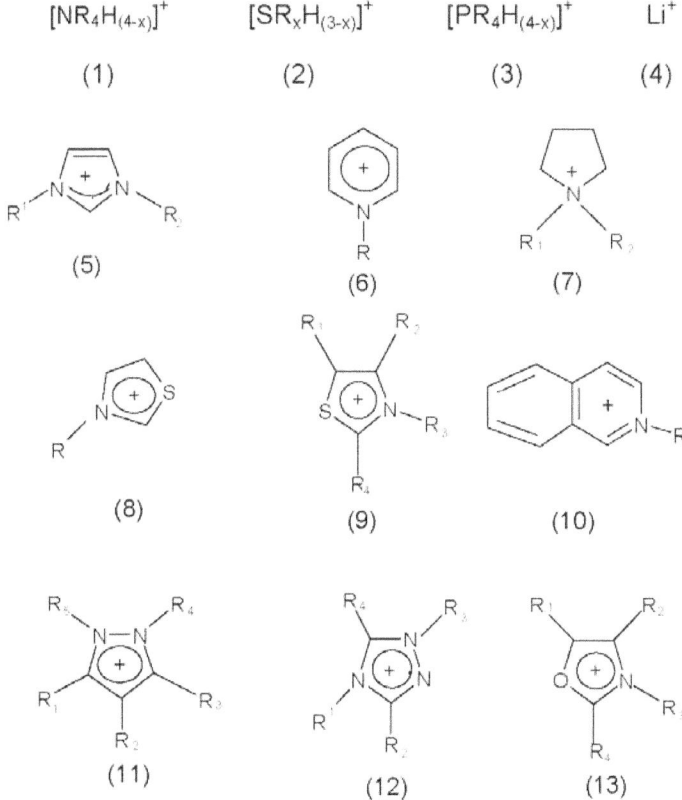

Figure 7. Examples of cation of ILs. Different substitutes of (1) ammonium, (2) sulfonium, (3) phosphonium, (4) lithium, (5) imidazolium, (6) pyridinium, (7) pyrrolidinium, (8) and (9) thiazonium, (10) isoquinolinium, (11) prazolium, (12) trizolium, and (13) oxazolium.

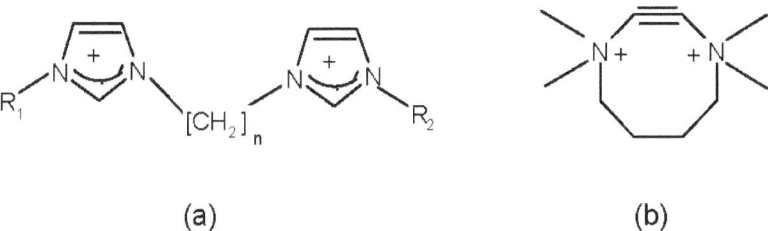

Figure 8. Some examples of poly-cations.

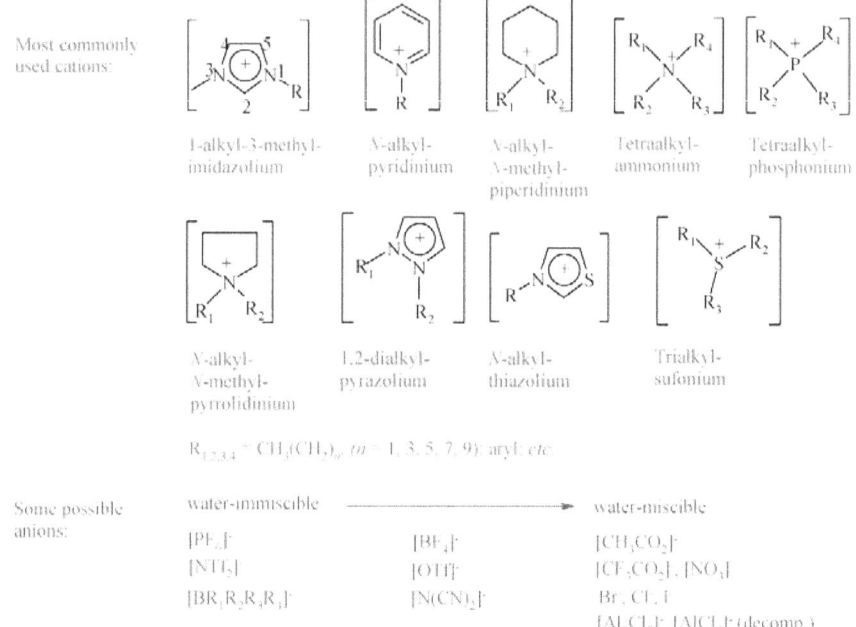

Most commonly used cations:

1-alkyl-3-methyl-imidazolium   N-alkyl-pyridinium   N-alkyl-N-methyl-piperidinium   Tetraalkyl-ammonium   Tetraalkyl-phosphonium

N-alkyl-N-methyl-pyrrolidinium   1,2-dialkyl-pyrazolium   N-alkyl-thiazolium   Trialkyl-sulfonium

$R_{1,2,3,4} = CH_3(CH_2)_n$, $n = 1, 3, 5, 7, 9$; aryl; etc.

Some possible anions:

water-immiscible   ⟶   water-miscible

$[PF_6]^-$ $[BF_4]^-$ $[CH_3CO_2]^-$
$[NTf_2]^-$ $[OTf]^-$ $[CF_3CO_2]^-$, $[NO_3]^-$
$[BR_2R_2R_3R_4]^-$ $[N(CN)_2]^-$ Br^-, Cl^-, I^-
   $[Al_2Cl_7]^-$, $[AlCl_4]^-$ (decomp.)

Figure 9. Most commonly used cations and anions.

3.2.1. ILs in catalysis

ILs are used as highly effective catalysts in a large number of industrial catalytic reactions [150]. As the functionalities of the ILs can be controlled and properly tuned as required for the reaction, a number of ILs were designed as substitutes for functional catalytic materials [151]. In terms of functionalities, the ILs can be classified as acidic, basic, bi-functional, task-specific, etc. Acidic ILs were reported efficiently for the alkylation of aromatics like phenols, cresols, catechols, cumenes, naphthalenes; biomass conversion, hydration reaction, ester hydrolysis, etc [152-173]. Basic ILs were reported as a substitute for a number of base-catalysed reactions like aldol condensation, esterification [174-175]. Bifunctional ILs were reported as important catalyst for a number of important reactions like oligomerization of olefins, esterification, hydroamination reaction, etc [176-178].

3.2.2. ILs as solvents

ILs are used efficiently as a solution to numerous separation problems. A number of complicated syntheses, as well as catalytic reactions, were successfully conducted using ILs as green solvents [179]. The volatile organic solvents can be substituted with nonvolatile ILs, which in turn, take care of the separation steps in industrial processes. As ILs do not vaporize, many contaminants can be easily removed from the IL solvents under high vacuum. A number of different inorganic or organic materials can be brought into same phases using IL solvents as they have a very long range of solubility [149]. As ILs are composed of poorly co-ordinated ions, they can be highly polar as well as noncoordinating at the same time. Several C-C coupling reactions were reported using ILs as solvents where the active metals like Pd, Pt, Ir, etc. were stabilized in the solvent [179]. Further, the use of ILs as extracting solvent has been widely studied utilizing its controllable polarity and solubility [149,180-182]. ILs with controlled polarity were used successfully to separate the mixture of aliphatic and aromatic hydrocarbon components like toluene, heptane, cyclohexane, xylene, etc.; useful bioactive components from plants. Also, the selective extraction of a number of critical and rare earth metals was performed using ILs as solvents [183-186]. ILs were also employed as a useful solvent in polymerization reactions [187,188]. ILs were also successfully used as solvents in controlling the selectivity as well as outcome of the Diels-Alder reaction [189-191]. Lewis acid catalyzed Beckmann rearrangement reaction were also reported using ILs as useful solvents [192]. Base-catalyzed reactions like aldol condensation, keto-enol tautomerization, nucleophilic substitution, esterification, etc. were also reported using ILs as solvents [193-202].

3.3. Industrial development of ionic liquid technologies

One of the successful implications of ionic liquids in industrial process is the Biphasic Acid Scavenging utilizing Ionic Liquids process abbreviated as BASIL™ process that was honoured with ECN Innovation Award in 2004. In this process, 1-methylimidazole is used to produce 1-methylimidazolium chloride ionic liquid that can be recovered as a separate phase from the reaction mixture. This improved process used smaller reactor and yielded 98% at space time of 690,000 kg m^{-3} h^{-1} as compared to the earlier used process that yielded only 50% at space time of only 8 kg m^{-3} h^{-1} [203]. Ionic liquids working as a separating agent or enhancer are used to break azeotropes such as water–tetrahydrofuran, water–chloroform, water–ethanol, etc. This helps in the cost reduction of entrainer isolation and recycle. BASF established the phosgene substitution by using HCl in ionic liquids and achieved product selectivity of 98%. BASF also made significant progress in the process of utilizing most plentiful renewable carbon resource, i.e. cellulose, which has the production and consumption rate of 7 x 10^{10} tons per year.

The first ionic liquid based pilot plant was operated by IFP (French Institute of Petroleum) in the Dimersol process in which high valued branched products like hexenes and octenes are produced from dimerization of alkenes especially propene (Dimersol-G) and butenes (Dimersol-X). This process is operated worldwide in around 35 plants where the turnover ranges from 20,000-90,000 tonnes/year in each plant. SASOL Ltd. Company established in South Africa has been rigorously exploiting the ionic liquids in metathesis and trimerization reactions, subjecting to Nobel Prize of 2005 in Chemistry. IoLiTech founded in 2003 is a pioneer provider of around 300 different high quality and purity ionic liquids for their commercial utilization as lubricants, solvents, and electrolytes. In 2009, IoLiTech expanded its sale to US, Canada and Mexico.

M/s BP and M/s ExxonMobil have been highly engaged in ionic liquid technology as evident from their frequent patent suite and articles. M/s BP is researching on the aromatics alkylation reaction using chloroaluminate ionic liquids. M/s Chevron in USA has filed several patents in the areas of petrochemical alkylation, emulsification, oligomerization, hydrotreatment of alkenes, and CO_2 removal in gas streams. PetroChina has developed a pilot scale process and named it 'Ionikylation'. This process employs ionic liquids for sulfuric acid alkylation giving a turnover of 65,000 tonne/year. Linde, France have fabricated a device called 'ionic compressor' by taking the advantage of low compressibility of ionic liquids and hence, using them as a liquid piston. This device is mainly used for gas compression at isothermal condition using a constant gas pressure of ~250 bar through supplying 500 m^3/h natural gas [203].

Honeywell UOP and Chevron U.S.A. Inc. collaborated in 2016 to invent a liquid alkylation technology and termed it 'ISOALKY' [204]. This technology replaces the conventional volatile toxic acids with strong acidic non-volatile ionic liquids for the production of alkylates. These alkylates are essential in refineries for high octane gasoline range fuel production from paraffins or olefins conversion. Sinochem Hongrun Petrochemical Co., Ltd. was the first Chinese manufacturer to license the Honeywell's ISOALKY technology for motor fuel production [204]. Liu et al. reported the scaling up of iso-butane/butene alkylation in ILs to produce high quality gasoline [205]. They improved the IL alkylation performance by adding a very small amount of benzene and observed the underlying mechanism by molecular dynamic simulations. They found that the small amount of benzene collects on the IL-hydrocarbon interfaces during the reaction which shielded the highly acidic chloroaluminate anions and thus, tuned acidity as well as created a channel for reactant transport to the catalyst. Singhal et al. reported a literature review on the alkene/iso-alkane reaction using ionic liquid catalysts. They described that in the presence of additives, strong acids, or promoters, ILs increases the butene conversion. The cationic and anionic species of ILs determine its acidity [206].

3.4. Alkylation reactions of aromatics

The aromatics generally phenols react with an alkylating agent form alkylphenols (C_1 to C_{12}) whose production rate is ~500,000 tonnes/year [207]. Alkylphenols are widely used as precursors of insecticides, herbicides, gasoline and polymer additives, dyes, lubricants, surfactants, synthetic resins, varnishes, and antioxidants. Alkylphenols require acidic catalysts for synthesis like phosphoric acid, sulfonic acids, anhydrous aluminum or zinc halides, boron trifluoride, etc [80-85]. Some other processes for particular alkylphenols involve alkylbenzene hydroxylation, alkylcyclohexanol

22

dehydrogenation, or acyclic ring closure. Among such routes, Friedel-Crafts alkylation is the most promising route for alkylphenol formation.

Most of the reports include investigations on solid acid catalysts. The problems related to the conventional solid acid catalysts are summarized in **Table 1**. A very little work has been reported on butylation of phenols using ionic liquids as catalysts. Zhao *et al*. [208] reported different ionic liquids which can catalyze the butylation reactions and are listed in **Table 2**. These catalysts have some activity toward the alkylation. The first two are Lewis acid type can be used when butylation agent is butene (olefins), and rests were used with *tert*-butyl alcohol as butylation agent.

3.4.1. *ILs in synthesis of alkylated aromatics*

A number of alkylation reactions have been reported till date using ILs as catalysts as well as solvents. Different functionalities like Bronsted acidity, Lewis acidity and Bronsted/Lewis dual acidity were explored for efficient synthesis of alkylated phenols as shown in **Figure 10**.

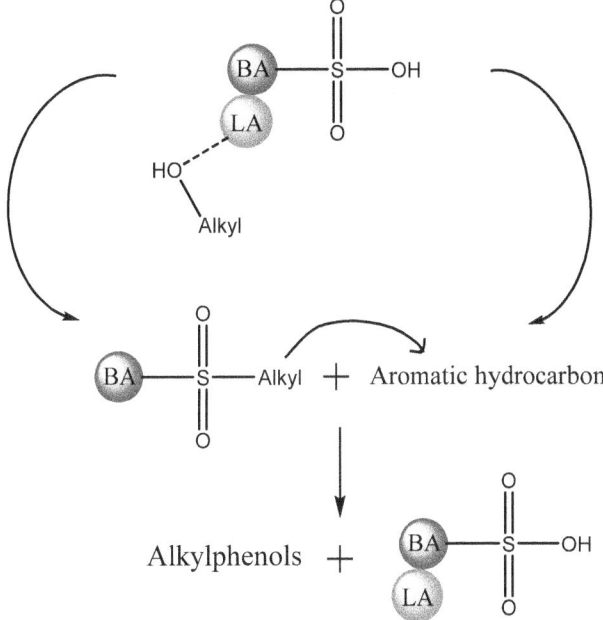

Figure 10. Alkylation reaction in presence of Bronsted-Lewis acidic IL.

3.4.2 *ILs as solvents in synthesis of alkylated aromatics*

In 2003, Shen *et al*. reported [BMIM]PF₆ ILs as solvents for the synthesis of alkylated phenols with TBA with 90% phenol conversion and 75% dialkylated compounds selectivity at a temperature of 60°C [160]. Further, in 2004, Shen *et al*. reported similar type of ILs in combination with mineral and solid acids as catalysts in the synthesis of *tert*butylated phenols [161]. The combination of the

ILs and H_3PO_4 resulted in 77.3% phenol conversion and 64.9% selectivity to the dialkylated compounds. The behavior of ILs as solvents in this alkylation reaction resulted in increased catalytic activity of the acidic catalysts.

3.4.3. Bronsted acidic ILs in synthesis of alkylated aromatics

Liu *et al.* reported $-SO_3H$ functionalized pyridinium based ILs for *p*-cresol alkylation giving its conversion of 79% conversion and 92% selectivity to monoalkylated products [209]. Elavarasan *et al.* reported $-SO_3H$ functionalized triethylammonium based Bronsted acidic ILs resulting in 82% conversion of phenol [210]. Kondamudi *et al.* reported the similar type of ILs for the synthesis of *tert*butylated *p*-cresols resulting in 84% *p*-cresol conversion and 72% selectivity towards monoalkylated cresols [163]. Bao *et al.* investigated multiple $-SO_3H$ functionalized imidazolium-based Bronsted acidic ILs for the synthesis of *tert*-butylated *p*-cresols with 85.3% conversion and 95.2% selectivity to monoalkylated product [165].

3.4.4. Lewis acidic ILs in synthesis of alkylated aromatics

Qiao *et al.* reported [BMIM]$AlCl_4$ type of Lewis acidic ILs for benzene alkylation with 1-dodecene. The use of the Lewis acidic ILs, in combination with HCl, resulted in 99.9% conversion of benzene and 41.2% selectivity to the *ortho*-alkylated products [153]. The excellent catalytic activity of these ILs was credited to superacidity induced by the existence of mineral Bronsted acids. Hence, a simultaneous presence of Bronsted and Lewis acidity can be considered to be an excellent catalyst combination for these type of reactions.

3.4.5. Bronsted-Lewis acidic ILs in synthesis of alkylated aromatics

Recently, Liu *et al.* reported Bronsted-Lewis acidic ILs, [HO$_3$SC$_3$NEt$_3$]Cl-ZnCl$_2$ for the alkylation of isobutane/isobutene system [211]. The synergistic effect of the Bronsted and Lewis acidity resulted in 91.7% yield of the C_8-alkylated product. Reported ILs were highly recyclable upto 10 cycles without any significant loss in catalytic activity.

Table 1. Summary of disadvantages of conventional and solid acid catalysts in butylation of phenols.

S.No	Catalyst	Problem with their use
1	$AlCl_3$, $FeCl_3$, $ZnCl_2$, HF,H_2SO_4,H_3PO_4	High corrosiveness, tedious work-up, undesirable side products, they are not environment friendly, difficult to separate from homogeneous mixture, requirement of stoichometric quantities etc.
2	Sulfated Zicronia	Unstable, probability of volatile sulfur compound generation while regeneration
3	Cation-exchange resins	Poor stability at higher temperatures.

4	Tungstophosphoric acid Supported on MCM-41	Poor selectivity, high pressure process (10MPa), coke formation
5	SiO_2-Al_2O_3, γ-Al_2O_3 and zeolites	Performance limited by diffusional constraints
6	Al-MCM-41 (Molecular Sieves)	Poor recyclability, selectivity and conversion decreases with each recycle.
8	WO_x/ZrO_2	More sensitive to reaction conditions, easy leaching from the support due to lack of stability

Table 2. Ionic liquids used in butylation reactions.

S. NO	Ionic liquid molecular formula	Ionic liquid Name
1	[emim]Cl/AlCl$_3$	1-butyl-3-ethyl imidazolium chloride
2	[amim]PF$_6$	1-alkyl-3-methyl imidazolium hexafluoro phosphate (alkyl group C$_4$-C$_8$)
3	[bmim]BF$_4$	1-butyl-3-methyl imidazolium tetrafloro borate
4	[bmim]HSO$_4$	1-butyl-3-methyl imidazolium Hydrogen sulfate
5	[(CH$_3$)$_3$N(CH$_2$)$_4$SO$_3$H][HSO$_4$]	N-(4-sulfonic acid) butyl trimethyl ammonium hydrogen sulfate
6	[(CH$_3$)$_3$N(CH$_2$)$_4$SO$_3$H][CH$_3$PhSO$_3$]	N-(4-sulfonic acid) butyl trimethyl ammonium toluene sulfonate
7	[(CH$_3$CH$_2$)$_3$N(CH$_2$)$_4$SO$_3$H][HSO$_4$]	N-(4-sulfonic acid) butyl triethyl ammonium hydrogen sulfate
8	[2-MPyH]Tfa	2-methyl puridinium trifluoro acetate
9	[2-MPyH]OTf	2-methyl puridinium trifluoro methane sulfonate
10	[2-MPyH][CH$_3$SO$_3$H]	2-methyl puridinium methane sulfonate
11	[Py (CH$_2$)$_4$SO$_3$H][HSO$_4$]	1-(4-sulfonic acid) butyl pyridinium hydrogen sulfate

4. CONCLUSIONS

Alkylation of aromatics with alcohols and transalkylation reactions are industrially important reactions resulting in the formation of gasoline additives and value-added chemicals. The widely used catalysts for these reactions are solid acid catalysts, which is recently being substituted by greener, highly efficient ionic liquids. These ionic liquids are efficient for their utilization both as solvents and acidic catalysts in the reaction. First, a review of the alkylation/transalkylation reactions over solid acid catalysts was discussed here followed by the more recent developments of ionic liquids. Different classes of ionic liquids were classified based on the ions (cations and anions) in their structure resulting in either Bronsted, Lewis or dual Bronsted-Lewis acidity. The industrial development of the ionic liquid technologies for alkylation reactions was also discussed. However, ionic liquids in transalkylation reactions is not yet explored and needs to be investigated thoroughly in the furture. Overall, this review could be an important asset for a quick view of the developments of alkylation and transalkylation technologies and the catalysts used over the past 30 years, for better task-specific catalyst design and reaction conditions in future for improved selective yields of the desired alkylation products and finally for commercialization of these technologies.

ACKNOWLEDGMENTS

The authors acknowledge IIT Delhi for providing research facilities.

REFERENCES

[1]. F. C. Friedel and J. M. Crafts, *Bull. Soc. Chim. Fr.*, **1877**, 27(2), 530.

[2]. R. M. Roberts and A. A. Khalaf, "Friedel-Crafts Alkylation Chemistry, A Century of Discovery", *Marcel Dekker Inc., New York*, **1982**.

[3]. F. A. Drahowzal, "Friedel-Crafts and Related Reactions", *Wiley-Interscience, New York*, **1964**, 2(1), 446.

[4]. S. H. Patkin and B.S. Friedmann, "Friedel-Crafts and Related Reactions", *Wiley-Interscience, New York*, **1964**, 2(1), 1.

[5]. C. L. Thomas, "Catalytic Processes and Proven catalysts", *Academic Press, New York*, **1988**.

[6]. G. R. Meima, M. J. M. Aalst, M. S. U. Samson, J. M. Garces, J. G. Lee, *Proc. Int. Zeolite Conf.* 9th Meeting, 1992, Eds. R. Von Ballmoos, J. B. Higgins, M. M. J. Treacy, Butterworth-Heinmann: Boston, Mass. **1993**, 2, 327-334.

[7]. A. R. Pradhan, B. S. Rao, V. P. Shiralkar, *J. Catal.* **1991**, 132, 79-84.

[8]. K. S. N. Reddy, B. S. Rao, V. P. Shiralkar, *Appl. Catal. A,* **1993**, 95, 53.

[9]. C. Perego, G. Pazzuconi, G. Girotti, G. Terzoni, *Eur. Pat. Appl. EP629599 A1* 21 Dec **1994**, 6pp.

[10]. W. W. Kaeding, C. Chu, L. B. Young, B. Weinstein, S. A. Butter, *J. Catal.* **1981**, 67, 159-174.

[11]. J. R. Anderson, K. Foger, T. Mole, R.A. Rajadhyaksha, J.V. Sanders, *J. Catal.* **1979**, 58, 114.

[12]. H.Y. Shen, Z.M.A. Judeh, C.B. Ching, Q.H. Xia, *Journal of Molecular Catalysis A: Chemical*, **2004**, 212, 301-308.

[13]. R. D. Kirk and D. F. Othmer, "Encyclopedia of Chemical Technology, 3 ed.", *Wiley Interscience, New York*, **1978**.

[14]. J. Tejero, F. Cunill, S. Manzano, *Applied Catalysis*, **1988**, 38, 327-340.

[15]. A. S. B. Schleppinghoff, H. L. Niederberger, H. V. Scheef, J. Grub, *Chem. Abstr*, **1991**.

[16]. F. Cunill, J. Tejero, J. F. Izquierdo, *Applied Catalysis*, **1987**, 34, 341-351.

[17]. N. Y. Chen, W. E. Garwood, F. G. Dwyer, "Shape Selective Catalysis in Industrial Applications", *Marcel Dekker Inc. New York*, **1989**.

[18]. J. E. Germain, "Catalytic Conversions of Hydrocarbons", *Academic Press, New York*, **1969**, 145.

[19]. H. Pines and O.T. Anigo, *J. Am. Chem. Soc.* **1958**, 80, 279.

[20]. T. Tecza, J. Lelakowska, Z. Lisicki, L. Franck, R. Czeniewski, R. Galbfach, M. Kozlowski, W. Redzenski, A. Rosciszewski, *PL 126095 B1*, **1983**, pp3.

[21]. L. H. Slaugh, *US 4375574 A1* March **1983**, pp4.

[22]. G. Messina, M. D. Moretti, G. Brundu, *Eur. Pat. EP 165215 A2* 18 Dec. **1985**, pp2.

[23]. S. T. Bakas, P. T. Barger, *M/s UOP Inc., USA, US 4870222 A* 26 Sep **1989**, pp14.

[24]. R. P. Arganbright, D. Hearn, *US 4950834 A* 21 August **1990**, pp9.

[25]. J. G. S. Lee, J. M. Garces, M. J. M. Vander Aalst, *EP 433932 A1* 26 June **1991**.

[26]. D. Mraree, A. Schnierer, D. Riecanova, J. Herain, V. Macho. *Collect. Czech. Chem. Commun.*, **1992**, 57(4), 896-900.

[27]. A. R. Pradhan and B.S. Rao, *Appl. Catal. A*, **1993**, 106(1), 143-153.

[28]. D. Olson and A. Brisio Eds. *Proceedings of the 6th International Zeolite Conference*, Butterworths, Surrey, United Kingdom, **1984**.

[29]. W. H. Flank. Ed. Am. *Chem. Soc. Simp. Series* No. **1980,** 135.

[30]. D. W. Breck, "Zeolite Molecular Sieves", *Wiley, New York*, **1974**.

[31]. R. Szostak, "Molecular Sieves: Principles of Synthesis and Identification", *Van Nostrand-Reinhold, New York*, **1989**.

[32]. W. O. Haag, R. M. Lobo, P. B. Weisz, *Nature*, **1984**, 309.

[33]. D. Barthomeuf, "Molecular Sieves-II, Ed. Kotzer, *American Chemical Society, Washington D.C.*, **1977**, 453.

[34]. S. Guo, Y. Wu, T. Jin, H. Wang, C. Dong, J. Zhang, M. Ding. *Fuel,* **2020**, 275, 117890.

[35]. S. T. Wilson, B. M. Lok, C. A. Messina, T. R. Cannan, E. M. Flanigen, *J. Amer. Chem. Soc.*, **1982**, 104, 1146.

[36]. J. M. Bennett, W. J. Dytrych, J. J. Pluth, J. V. Smith, *Zeolites*, **1986**, 6, 349.

[37]. B. M. Lok, C. A. Messina, R. L. Patton, T. R. Cannan, E. M. Flanigen, *US Patent 4440871*, **1984**.

[38]. E. M. Flanigen, B. M. Lok, R. L. Patton, S. T. Wilson, "New Developments in Zeolite Science and Technology", Eds. Y. Murakami, A. Iijima and J. W. Ward, *Elsevier, New York, NY,* **1986**, 103-112.

[39]. J. W. Ward, "Zeolite Chemistry and Catalysis", Ed. J.A. Rabo, *ACS Monograph, Washington D.C*, **1976**, 171, 226.

[40]. S. G. Hegde, P. Ratnaswamy, L. M. Kustov, V. B. Kazansky, *Zeolites*, **1988**, 8, 137.

[41]. M. Briend, A. Lancy, S. Dzwigaj, D. Barthomeuf, *Stud. Surf. Sci. Catal.* **1991**, 69, 313.

[42]. S. W. Kaiser, *US patent US 4524234*, **1985**, pp10.

[43]. R. J. Pellet, G. N. Long, J. A. Rabo, "New Developments in Zeolite Science and Technolgy*"*, *Proceedings of the 7 th International Zeolite conference*, Eds. J. Murakami, A. lijima and J.W. Ward, *Elsevier, New York*, **1986**, pp843-849.

[44]. J. A. Martens, P.J. Grobet, P.A. Jacobs, *J. Catal.* **1990**, 126, 299.

[45]. R. J. Pelter, P.K. Coughlin, E. S. Shashoum, J. A. Rabo, "Perspectives in Molecular Sieve Science", Eds. W.H. Flank, T. E. Whyte, *ACS Symposium Series* **1988**, 368, 512-531.

[46]. E. M. Flanigen, R. L. Patton, S.T. Wilson, "Innovations in Zeolite Science", Eds. P. J. Grobet, W. J. Mortier, E.F. Vasant and G. Schulz- Eckloff, *Elsevier, Amsterdam*, **1988**, 13-28.

[47]. N. J. Tapp, N. B. Milestone, D. M. Bibby, "Innovations in Zeolite Science", Eds. P. J. Grobet, W. J. Mortier, E.F. Vasant and G. Schulz- Eckloff, *Elsevier, Amsterdam*, **1988**, 402.

[48]. C. A. Sears, *The J. of Organic Chem.*, **1948**, 13, 120-122.

[49]. M. M. Sharma, *Reactive and Functional Polymers*, **1995**, 26, 3-23.

[50]. Z. R. Su and T. J. Wang, *Reactive and Functional Polymers*, **1995**, 28, 97-102.

[51]. M. A. Harmer and Q. Sun, *Applied Catalysis A: General*, **2001**, 221, 45-62.

[52]. C. B. Campbell, A. Onopchenko, D. C. Young, *Industrial & Engineering Chemistry Research*, **1990**, 29, 642-647.

[53]. E. Dumitriu and V. Hulea, *Journal of Catalysis*, **2003**, 218, 249-257.

[54]. K. Ojha, N. C. Pradhan, A. N. Samanta, *Chemical Engineering Journal*, **2005**, 112, 109-115.

[55]. A. Sakthivel, S. K. Badamali, P. Selvam, *Microporous and Mesoporous Materials*, **2000**, 39, 457-463.

[56]. N. Bhatt, P. Sharma, A. Patel, P. Selvam, *Catalysis Communications*, **2008**, 9, 1545-1550.

[57]. K. Chandler, C. L. Liotta, C. A. Eckert, D. Schiraldi, *AIChE Journal*, **1998**, 44, 2080-2087.

[58]. J. K. Gehlawat and M. M. Sharma, *J. appl. Chem.*, **1970**, 20, 93-98.

[59]. K. G. Chandra, M. M. Sharma, *Catalysis Letters*, **1993**, 19, 309-317.

[60]. A. H. Padmasri, V. D. Kumari, P. K. Rao, *Studies in Surface Science and Catalysis*, **1998**, 563-571.

[61]. K. Zhang, C. Huang, H. Zhang, S. Xiang, S. Liu, D. Xu, H. Li, *Applied Catalysis A: General*, **1998**, 166, 89-95.

[62]. C. V. Satyanarayana, S. Upadhyayula, B. S. Rao, *Studies in Surface Science and Catalysis*, **2001**, 238-238.

[63]. K. Zhang, H. Zhang, G. Xu, S. Xiang, D. Xu, S. Liu, H. Li, *Applied Catalysis A: General*, **2001**, 207, 183-190.

[64]. X. H. Tang, X. L. Fu, H. Y. Jiang, *Studies in Surface Science and Catalysis*, **2002**, 525-530.

[65]. G. D. Yadav, N. S. Doshi, *Applied Catalysis A: General*, **2002**, 236, 129-147.

[66]. A. H. Padmasri, A. Venugopal, V. D. Kumari, K. S. R. Rao, P. K. Rao, *Journal of Molecular Catalysis A: Chemical*, **2002**, 188, 255-265.

[67]. S. E. Dapurkar, P. Selvam, *Applied Catalysis A: General*, **2003**, 254, 239-249.

[68]. X. Tang, A. Zhang, J. Liu, *Studies in Surface Science and Catalysis*, **2004**, 2754-2759.

[69]. T. Mathew, B. S. Rao, C. S. Gopinath, *Journal of Catalysis*, **2004**, 222, 107-116.

[70]. B. M. Devassy, G. V. Shanbhag, F. Lefebvre, S. B. Halligudi, *Journal of Molecular Catalysis A: Chemical*, **2004**, 210, 125-130.

[71]. M. Karthik, A. Vinu, A. K. Tripathi, N. M. Gupta, M. Palanichamy, V. Murugesan, *Microporous and Mesoporous Materials*, **2004**, 70, 15-25.

[72]. A. Vinu, K. U. Nandhini, V. Murugesan, W. Böhlmann, V. Umamaheswari, A. Pöppl, M. Hartmann, *Applied Catalysis A: General*, **2004**, 265, 1-10.

[73]. A. B. Shinde, N. B. Shrigadi, S. D. Samant, *Applied Catalysis A: General*, **2004**, 276, 5-8.

[74]. K. U. Nandhini, B. Arabindoo, M. Palanichamy, V. Murugesan, *Journal of Molecular Catalysis A: Chemical*, **2004**, 223, 201-210.

[75]. E. Dumitriu, D. Meloni, R. Monaci, V. Solinas, *Comptes Rendus Chimie*, **2005**, 8, 441-456.

[76]. A. Vinu, B. M. Devassy, S. B. Halligudi, W. Böhlmann, M. Hartmann, *Applied Catalysis A: General*, **2005**, 281, 207-213.

[77]. B. M. Devassy, G. V. Shanbhag, S. P. Mirajkar, W. Böhringer, J. Fletcher, S. B. Halligudi, *Journal of Molecular Catalysis A: Chemical*, **2005**, 233, 141-146.

[78]. S. Sarish, B. M. Devassy, S. B. Halligudi, *Journal of Molecular Catalysis A: Chemical*, **2005**, 235, 44-51.

[79]. S. Wu, J. Huang, T. Wu, K. Song, H. Wang, L. Xing, H. Xu, L. Xu, J. Guan, Q. Kan, *Chinese Journal of Catalysis*, **2006**, 27, 9-14.

[80]. G. D. Yadav and G. S. Pathre, *Microporous and Mesoporous Materials*, **2006**, 89, 16-24.

[81]. B. M. Devassy, G. V. Shanbhag, S. B. Halligudi, *Journal of Molecular Catalysis A: Chemical*, **2006**, 247, 162-170.

[82]. Z. Duan, Y. Gu, J. Zhang, L. Zhu, Y. Deng, *Journal of Molecular Catalysis A: Chemical*, **2006**, 250, 163-168.

[83]. M. Kurian and S. Sugunan, *Catalysis Communications*, **2006**, 7, 417-421.

[84]. G. Kamalakar, K. Komura, Y. Sugi, *Applied Catalysis A: General*, **2006**, 310, 155-163.

[85]. S. M. Kumbar, G. V. Shanbhag, F. Lefebvre, S. B. Halligudi, *Journal of Molecular Catalysis A: Chemical*, **2006**, 256, 324-334.

[86]. K. U. Nandhini, J. H. Mabel, B. Arabindoo, M. Palanichamy, V. Murugesan, *Microporous and Mesoporous Materials*, **2006**, 96, 21-28.

[87]. G. S. Kumar, M. Vishnuvarthan, M. Palanichamy, V. Murugesan, *Journal of Molecular Catalysis A: Chemical*, **2006**, 260, 49-55.

[88]. X. H. Tang, A. P. Zhang, J. Liu, X. L. Fu, *Studies in Surface Science and Catalysis*, **2007**, 1454-1459.

[89]. M. Selvaraj and S. Kawi, *Microporous and Mesoporous Materials*, **2007**, 98, 143-149.

[90]. M. Selvaraj and P. K. Sinha, *Journal of Molecular Catalysis A: Chemical*, **2007**, 264, 44-49.

[91]. E. P. Ng, H. Nur, K. L. Wong, M. N. M. Muhid, H. Hamdan, *Applied Catalysis A: General*, **2007**, 323, 58-65.

[92]. G. S. Kumar, S. Saravanamurugan, M. Hartmann, M. Palanichamy, V. Murugesan, *Journal of Molecular Catalysis A: Chemical*, **2007**, 272, 38-44.

[93]. B. M. Reddy, M. K. Patil, G. K. Reddy, B. T. Reddy, K. N. Rao, *Applied Catalysis A: General*, **2007**, 332, 183-191.

[94]. D. Meloni, E. Dumitriu, R. Monaci, V. Solinas, *Studies in Surface Science and Catalysis*, **2008**, 1111-1114.

[95]. L. Xu, S. Wu, J. Guan, H. Wang, Y. Ma, K. Song, H. Xu, H. Xing, C. Xu, Z. Wang, Q. Kan, *Catalysis Communications*, **2008**, 9, 1272-1276.

[96]. S. V. Priya, J. H. Mabel, S. Gopalakrishnan, M. Palanichamy, V. Murugesan, *Journal of Molecular Catalysis A: Chemical*, **2008**, 290, 60-66.

[97]. K. Bachari, A. Touileb, M. Touati, O. Cherifi, *Journal of Molecular Catalysis A: Chemical*, **2008**, 294, 61-67.

[98]. E. Modrogan, M. H. Valkenberg, W. F. Hoelderich, *Journal of Catalysis*, **2009**, 261, 177-187.

[99]. X. Liao, G. Chen, G. Liu, L. Sun, W. Huo, W. Zhang, M. Jia, *Microporous and Mesoporous Materials*, **2009**, 124, 210-217.

[100]. J. Zhou, X. Liu, S. Zhang, J. Mao, X. Guo, *Catalysis Today*, **2010**, 149, 232-237.

[101]. M. Ghiaci and B. Aghabarari, *Chinese Journal of Catalysis*, **2010**, 31, 759-764.

[102]. L. F. Albright, *Chemtech*, **1998**, 28, 46-53.

[103]. A. Feller, I. Zuazo, A. Guzman, J. O. Barth, J. A. Lercher, *Journal of Catalysis*, **2003**, 216, 313-323.

[104]. P. Rao, S. R. Vatcha, *Oil Gas J.*, **1996**, 94, 56.

[105]. A. Corma, A. Martínez, C. Martínez, *Journal of Catalysis*, **1994**, 146, 185-192.

[106]. A. Corma, A. Martinez, C. Martinez, *Journal of Catalysis*, **1994**, 149, 52-60.

[107]. A. Corma, A. Martinez, C. Martinez, *Catalysis Letters*, **1994**, 28, 187-201.

[108]. M. G. Clerici, A. de Angelis, P. Ingallina, *Italian Patent*, **1995**.

[109]. J. D. Holbrey, *Clean Prod. Proc.*, **1999**, 1, 223-236.

[110]. A. J. Carmichael, M. J. Earle, J. D. Holbrey, P. B. McCormac, K. R. Seddon, *Organic Letters*, **1999**, 1, 997-1000.

[111]. Y. Chauvin, H. Y. Olivier-Bourbigou, *Chemtech*, **1995**, 25, 26-30.

[112]. Y. Chauvin, L. Mussman, H. Olivier, *Angew. Chem.-Internat. Ed. Eng.*, **1996**, 34, 2698.

[113]. R. F. deSouza, V. Rech, J. Dupont, *Advanced Synthesis & Catalysis*, **2002**, 344, 153-155.

[114]. P. J. Dyson, *Electrochem. Soc. Proc.*, **2000**, 99-41, 161.

[115]. B. Ellis, W. Keim, P. Wasserscheid, *Chemical Communications*, **1999**, 337-338.

[116]. M. Freemantle, *Chem. Eng. News*, **1998**, 76, 32-37.

[117]. N. Gathergood, M. T. Garcia, P. J. Scammells, *Green Chemistry*, **2004**, 6, 166-175.

[118]. R. Hart, P. Pollet, D. J. Hahne, E. John, V. Llopis-Mestre, V. Blasucci, H. Huttenhower, W. Leitner, C. A. Eckert, C. L. Liotta, *Tetrahedron*, **2010**, 66, 1082-1090.

[119]. J. Huang, A. Riisager, R. W. Berg, R. Fehrmann, *Journal of Molecular Catalysis A: Chemical*, **2008**, 279, 170-176.

[120]. P. J. Dyson, D. J. Ellis, T. Welton, D. G. Parker, *Chemical Communications*, **1999**, 25-26.

[121]. A. J. Carmichael, J. M. Crosthwaite, E. J. Maginn, J. F. Brennecke, *Journal of Physical Chemistry B*, **2006**, 110, 9354-9361.

[122]. K. S. Kim, S. Choi, D. Demberelnyamba, H. Lee, J. Oh, B. B. Lee, S. J. Mun, *Chemical Communications*, **2004**, 828-829.

[123]. M. A. Klingshirn, G. A. Broker, J. D. Holbrey, K. H. Shaughnessy, R. D. Rogers, *Chemical Communications*, **2002**, 1394-1395.

[124]. J. A. Laszlo and D. L. Compton, *Biotechnology and Bioengineering*, 2001, 75, 181-186.

[125]. P. R. Likhar, S. Roy, M. Roy, M. S. Subhas, M. L. Kantam, *Catalysis Communications*, **2009**, 10, 728-731.

[126]. D. R. MacFarlane, J. Golding, S. Forsyth, M. Forsyth, G. B. Deacon, *Chemical Communications*, **2001**, 1430-1431.

[127]. A. L. Monteiro, F. K. Zinn, R. F. deSouza, J. Dupont, *Tetrahedron: Asymmetry*, **1997**, 8, 177-179.

[128]. K. Qiao, H. Hagiwara, C. Yokoyama, *Journal of Molecular Catalysis A: Chemical*, **2006**, 246, 65-69.

[129]. D. Saha, A. Saha, B. C. Ranu, *Tetrahedron Letters*, **2009**, 50, 6088-6091.

[130]. K. R. Seddon and A. Stark, *Green Chemistry*, **2002**, 4, 119-123.

[131]. R. A. Sheldon, R. M. Lau, M. J. Sorgedrager, F. van Rantwijk, K. R. Seddon, *Green Chemistry*, **2002**, 4, 147-151.

[132]. C. E. Song, E. J. Roh, *Chemical Communications*, **2000**, 837-838.

[133]. P. A. Z. Suarez, J. E. L. Dullius, S. Einoft, R. F. deSouza, *Inorganica Chimica Acta*, **1997**, 255, 207.

[134]. R. Sugimura, K. Qiao, D. Tomida, C. Yokoyama, *Catalysis Communications*, **2007**, 8, 770-772.

[135]. P. Wasserscheid, T. Welton, *Outlook, Wiley-VCH Verlag GmbH & Co. KGaA*, **2003**.

[136]. T. Welton, *Chemical Reviews*, **1999**, 99, 2071-2084.

[137]. T. Welton, *Coordination Chemistry Reviews*, **2004**, 248, 2459-2477.

[138]. D. Zhao, M. Wu, Y. Kou, E. Min, *Catalysis Today*, **2002**, 74, 157-189.

[139]. Z. Zhao, Z. Li, G. Wang, W. Qiao, L. Cheng, *Applied Catalysis A: General*, **2004**, 262, 69-73.

[140]. H. Olivier-Bourbigou, L. Magna, *Journal of Molecular Catalysis A: Chemical*, **2002**, 182-183, 419-437.

[141]. P. Wasserscheid, T. Welton, *Wiley-VCH Verlag GmbH & Co. KGaA*, **2002**.

[142]. S. I. Lall, D. Mancheno, S. Castro, V. Behaj, J. I. Cohen, R. Engel, *Chemical Communications*, **2000**, 2413-2414.

[143]. R. S. Varma, V. V. Namboodiri, *Chemical Communications*, **2001**, 643-644.

[144]. G. Fitzwater, W. Geissler, R. Moulton, N. V. Plechkova, A. Robertson, K. R. Seddon, J. Swindall, K. W. Joo, *Ionic Liquids: Source of Innovation,* **2005**, *Report Q00*, 1.

[145]. J. P. Hallett, T. Welton, *Chem. Rev.* **2011**, 111(5), 3508.

[146]. P. Wasserscheid and W. Keim, *Angew. Chemie,* **2000**, 39(21), 3772.

[147]. A. Amareskara, *Chem. Rev.* **2016**, 116, 6133.

[148]. T. Welton, *Chem. Rev.* **1999**, 99 (8), 2071.

[149]. R. I. Canales, J. F. Brennecke, *J. Chem. Eng. Data* **2016**, 61 (5), 1685.

[150]. Q. Zhang, S. Zhang, Y. Deng, *Green Chem.* **2011**, 13 (10), 2619.

[151]. K. Goossens, K. Lava, C. W. Bielawski, K. Binnemans, *Chem. Rev.* **2016**, 116 (8), 4643.

[152]. L. Y. Piao, X. Fu, Y. L. Yang, G. H. Tao, Y. Kou, *Catal. Today,* **2004**, 93–95, 301.

[153]. C. Z. Qiao, Y. F. Zhang, J. C. Zhang, C. Y. Li, *Appl. Catal. A Gen.,* **2004**, 276 (1–2), 61.

[154]. K. Qiao, Y. Deng, *J. Mol. Catal. A Chem.* **2001**, 171 (1–2), 81.

[155]. J. Joni, D. Schmitt, P. S. Schulz, T. J. Lotz, P. Wasserscheid, *J. Catal.* **2008**, 258 (2), 401.

[156]. Z. Zhao, Z. Li, G. Wang, W. Qiao, L. Cheng, *Appl. Catal. A Gen.* **2004**, 262 (1), 69.

[157]. Z. K. Zhao, W. H. Qiao, Z. S. Li, G. R. Wang, L. B. Cheng, *J. Mol. Catal. A Chem.* **2004**, 222 (1–2), 207.

[158]. M. Chen, Y. Luo, G. Li, M. He, J. Xie, H. Li, X. Yuan, *Korean J. Chem. Eng.* **2009**, 26 (6), 1563.

[159]. J. A. Boon, J. A. Levisky, J. L. Pflug, J. S. Wilkes, *J. Org. Chem.* **1986**, 51 (4), 480.

[160]. H. Shen, Z. M. A. Judeh, C. B. Ching, **2003**, 44, 981.

[161]. H. Shen, Z. M. A. Judeh, C. B. Ching, Q. H. Xia, *J. Mol. Catal. A Chem.* **2004**, 212 (1–2), 301.

[162]. J. Gui, H. Ban, X. Cong, X. Zhang, Z. Hu, Z. Sun, *J. Mol. Catal. A Chem.* **2005**, 225 (1), 27.

[163]. K. Kondamudi, P. Elavarasan, P. J. Dyson, S. Upadhyayula, *J. Mol. Catal. A Chem.* **2010**, 321 (1–2), 34.

[164]. P. Elavarasan, K. Kondamudi, S. Upadhyayula, *World Acad. Science, Eng. Technol.* **2010**, 42, 177.

[165]. S. Bao, N. Quan, J. Zhang, J. Yang, *Chinese J. Chem. Eng.* **2011**, 19 (1), 64.

[166]. X. Liu, M. Liu, X. Guo, J. Zhou, *Catal. Commun.* **2008**, 9(1), 1.

[167]. H. Wang, G. Gurau, R. D. Rogers, *Chem. Soc. Rev.* **2012**, 41 (4), 1519.

[168]. Z. D. Ding, J. C. Shi, J. J. Xiao, W. X. Gu, C. G. Zheng, H. J. Wang, *Carbohydr. Polym.* **2012**, 90 (2), 792.

[169]. F. Parveen, T. Patra, S. Upadhyayula, *Carbohydr. Polym.* **2015**, 135, 280.

[170]. L. Zhou, R. Liang, Z. Ma, T. Wu, Y. Wu, *Bioresour. Technol.* **2013**, 129 (2013), 450.

[171]. R. Kore and R. Srivastava, *Tetrahedron Lett.* **2012**, 53 (26), 3245.

[172]. R. Kore and R. Srivastava, *J. Mol. Catal. A Chem.* **2013**, 376, 90.

[173]. D. Q. Nguyen, J. H. Oh, C. S. Kim, S. W. Kim, H. Kim, H. Lee, H. S. Kim, *Bull. Korean Chem. Soc.* **2007**, 28 (12), 2299.

[174]. C. P. Mehnert, N. C. Dispenziere, R. A. Cook, *Chem. Commun.* **2002**, 15, 1610.

[175]. S. A. Forsyth, D. R. MacFarlane, R. J. Thomson, M. von Itzstein, *Chem. Commun.* **2002**, 7, 714.

[176]. Y. Gu, F. Shi, Y. Deng, *Catal. Commun.* **2003**, 4 (11), 597.

[177]. X. X. Han, H. Du, C. T. Hung, L. L. Liu, P. H. Wu, D. H. Ren, S. J. Huang, S. B. Liu, *Green Chem.* **2014**, 17 (1), 499.

[178]. L. Yang, L. Xu, C. G. Xia, *Synthesis* **2009**, 12, 1969.

[179]. W. Kunz and K. Häckl, *Chem. Phys. Lett.* **2016**, 661, 6.

[180]. T. Sukhbaatar, S. Dourdain, R. Turgis, J. Rey, G. Arrachart, S. Pellet-Rostaing, *Chem. Commun.* **2015**, 51 (88), 15960.

[181]. B. Tang, W. Bi, M. Tian, K. H. Row, *J. Chromatogr. B Anal. Technol. Biomed. Life Sci.* **2012**, 904, 1.

[182]. *Green Solvents II*; Mohammad, A., Inamuddin, D., Eds.; Springer Netherlands: Dordrecht, **2012**.

[183]. L. Fischer, T. Falta, G. Koellensperger, A. Stojanovic, D. Kogelnig, M. Galanski, R. Krachler, B. K. Keppler, S. Hann, *Water Res.* **2011**, 45 (15), 4601.

[184]. U. Domańska and A. Rękawek, *J. Solution Chem.* **2009**, 38 (6), 739.

[185]. T. Makanyire, S. Sanchez-Segado, A. Jha, A. *Adv. Manuf.* **2016**, 4 (1), 33.

[186]. C. H. C. Janssen, N. A. Macías-Ruvalcaba, M. Aguilar-Martínez, M. N. Kobrak, *Int. Rev. Phys. Chem.* **2015**, 34 (4), 591.

[187]. P. Kubisa, *Prog. Polym. Sci.* **2004**, 29 (1), 3.

[188]. P. Kubisa, *Prog. Polym. Sci.* **2009**, 34 (12), 1333.

[189]. A. P. Abbott, G. Capper, D. L. Davies, R. K. Rasheed, *Green Chem.* **2002**, 4, 24.

[190]. D. Yin, C. Li, B. Li, L. Tao, D. Yin, *Adv. Synth. Catal.* **2005**, 347 (1), 137.

[191]. A. Kumar and S. S. Pawar, **2004**, 8, 1419.

[192]. A. Zicmanis, S. Katkevica, P. Mekss, *Catal. Commun.* **2009**, 10 (5), 614.

[193]. I. Cota, R. Gonzalez-Olmos, M. Iglesias, F. Medina, *J. Phys. Chem. B* **2007**, 111 (43), 12468.

[194]. V. K. Aggarwal, I. Emme, A. Mereu, *Chem. Commun.* **2002**, 15, 1612.

[195]. P. Kotrusz, I. Kmentová, B. Gotov, S. Toma, E. Solcániová, *Chem. Commun.* **2002**, 2, 2510.

[196]. P. Formentín, H. García, A. Leyva, *J. Mol. Catal. A Chem.* **2004**, 214 (1), 137.

[197]. G. Angelini, P. De Maria, C. Chiappe, A. Fontana, C. Gasbarri, G. Siani, *J. Org. Chem.* **2009**, 74 (17), 6572.

[198]. C. Chiappe and D. Pieraccini, *Green Chem.* **2003**, 5 (2), 193.

[199]. K. Sandhya and B. Ravindranath, *Tetrahedron Lett.* **2008**, 49 (15), 2435.

[200]. A. Natrajan and D. Wen, *Green Chem.* **2011**, 13 (4), 913.

[201]. A. R. Gholap, K. Venkatesan, T. Daniel, R. J. Lahoti, K. V. Srinivasan, *Green Chem.* **2003**, 5 (6), 693.

[202]. A. A. M. Lapis, L. F. De Oliveira, B. A. D. Neto, J. Dupont, *ChemSusChem* **2008**, 1 (8–9), 759.

[203]. N. V. Plechkova, K. R. Seddon, *Chemical Society Reviews*, **2008**, 37, 123-150.

[204]. M. P. Bailey, "Honeywell UOP licenses ionic-liquids alkylation technology in china", *Chemical Engineering Essentials for the CPI Professionals*, **2019**.

[205]. C. P. Huang, Z. C. Liu, C. M. Xu, B. H. Chen, Y. F. Liu. *App. Catal. A: General,* **2004**, 277, 41-43.

[206]. S.Singhal, S. Agarwal, M. Singh, S. Rana, S. Arora, N. Singhal. *J. of Molecular Liquids,* **2019**, 285, 299-313.

[207]. A. K. Venkatesan, R. U. Halden. *Environ Pollut.*, **2013**, 174, 189–193.

[208]. D. Zhao, M. Wu, Y. Kou, E. Min, *Catalysis Today*, **2002**, 74, 157-189.

[209]. X. Liu, J. Zhou, X. Guo, M. Liu, X. Ma, C. Song, C. Wang, V. Pennsyl, *Ind. Eng. Chem. Res.* **2008**, 47, 5298.

[210]. P. Elavarasan, K. Kondamudi, S. Upadhyayula, *Chem. Eng. J. J.* **2011**, *166*, 340.

[211]. S. Liu, C. Chen, F. Yu, L. Li, Z. Liu, S. Yu, C. Xie, F. Liu, *Fuel* **2015**, 159, 803.

Publisher: Eliva Press SRL

Email: info@elivapress.com

Eliva Press is an independent publishing house established for the publication and dissemination of academic works all over the world. Company provides high quality and professional service for all of our authors.

Our Services:
Free of charge, open-minded, eco-friendly, innovational.

-Free standard publishing services (manuscript review, step-by-step book preparation, publication, distribution, and marketing).
-No financial risk. The author is not obliged to pay any hidden fees for publication.
-Editors. Dedicated editors will assist step by step through the projects.
-Money paid to the author for every book sold. Up to 50% royalties guaranteed.
-ISBN (International Standard Book Number). We assign a unique ISBN to every Eliva Press book.
-Digital archive storage. Books will be available online for a long time. We don't need to have a stock of our titles. No unsold copies. Eliva Press uses environment friendly print on demand technology that limits the needs of publishing business. We care about environment and share these principles with our customers.
-Cover design. Cover art is designed by a professional designer.
-Worldwide distribution. We continue expanding our distribution channels to make sure that all readers have access to our books.

www.elivapress.com